Advances in Bulk Crystal and Thin Film Formation